Exploring The BUILDING BLOCKS of Science

Book 1

TEACHER'S MANUAL

REBECCA W. KELLER, PhD

Exploring the Building Blocks of Science Book 1 Teacher's Manual
ISBN 978-1-936114-32-0

Published by Gravitas Publications Inc.
www.gravitaspublications.com

A Note From the Author

This curriculum is designed to provide an introduction to the sciences for students in kindergarten through first grade. The *Building Blocks of Science Laboratory Notebook: Book 1* is intended to be the first step in developing a framework for real science. This teacher's manual will help you guide students through the laboratory experiments. The series of experiments in the *Laboratory Notebook* will help the students develop the skills needed for the first step in the scientific method — making good observations.

There are different sections in each chapter. The section called *Observe It* helps the students explore how to make good observations. The *Think About It* section provides questions for the students to think about and to use to make further observations. In every chapter there is a *What Did You Discover?* section that gives the students an opportunity to summarize the observations they have made. A section called *Why?* provides a short explanation of what students may or may not have observed. And finally, in each chapter there is a section called *Just For Fun* that contains an additional activity.

The experiments take up to 1 hour. The materials needed for each experiment are listed on the next page and also at the beginning of each experiment.

Enjoy!

Rebecca W. Keller, PhD

Materials at a Glance

Experiment 1	Experiment 3	Experiment 4	Experiment 5	Experiment 6
colored pencils night sky a living thing to observe (frog, ant, plant, or other)	magnifying glass household items such as: cotton balls rubber bands pencil several food items such as: crackers cheese marshmallow beans color-coated candy (such as M&Ms)	Legos marshmallows, small (1 pkg) marshmallows, large (1 pkg) toothpicks	4 or more clear plastic cups or glasses marking pen measuring cup measuring spoons food items-180 ml (3/4 cup) each: lemon juice vinegar milk water baking soda - 90 ml (6 Tbsp.) *Just For Fun* section: baking soda vinegar sugar -Or- 2 or more food items chosen by student	non-living object to observe (such as a rock or piece of wood) living thing to observe (such as an ant, frog, bird, cat, or dog) colored pencils
Experiment 2				
watercolor paints water in a container paintbrush paper to paint on, several sheets scissors tape				

Experiment 7	Experiment 8	Experiment 9	Experiment 10	Experiment 11
cotton balls rubber ball tennis ball banana apple rocks Legos other objects colored pencils	internet access and/or reference books colored pencils	milk, .25 l (1 cup) plain yogurt, .5 liter (2 cups) fork spoon cups or small bowls (several) food items such as honey, berries, chopped fruit or vegetables, spices, herbs, cocoa, chocolate chips, etc. (*Just For Fun* section)	2 tennis balls other objects such as: apple orange rubber ball cotton ball or feather	large marshmallow tennis ball objects such as: rubber ball lemon or lime rock banana pliers

Experiment 12	Experiment 13	Experiment 14	Experiment 15	Experiment 16
toy car stiff cardboard or board (approximately .3 meter wide x 1 meter long [1 foot x 3 feet]) marshmallows (several) **Optional** pennies tape	2 marbles 3 playing cards shallow jar top vinegar baking soda measuring spoons **Suggested** dominoes blocks electric car electric train marshmallow tongue depressor steel ball other objects chosen by the student	colored pencils	small shovel or garden trowel small pail or plastic container measuring cup dirt that contains rocks (.25 liter [1 cup]) glass container, tall clear (approx. size: .5 liter [2 cups]) flour (60 ml [1/4 cup]) water cake mix and items needed to make the cake nuts, gumdrops, chocolate chips, and/or M&Ms	baseball or similar hard-centered ball balloon water piece of string to tie balloon closed colored pencils **Optional** funnel

Experiment 17	Experiment 18	Experiment 19	Experiment 21	Experiment 22
2 liters (8 cups) or more of dirt suitable for making mud pies 1.75 liters (7 cups) or more of water 15 ml (1 Tbsp.) baking soda 15 ml (1 Tbsp.) vinegar measuring cup measuring spoon 3 containers for mixing mud (about 1.75 liter [7 cups] size) spoon or other implement for mixing mud garden trowel bucket paper marking pen pencil colored pencils	clear night sky colored pencils **Note:** This experiment will take 6 days to complete.	basketball ping-pong ball flashlight empty toilet paper tube glue or tape scissors marking pen a dark room **Experiment 20** colored pencils night sky **Note:** This experiment will take two weeks to complete.	8 styrofoam balls: 1 - 10 cm (4 in) 1 - 7.5 cm (3 in) 2 - 5 cm (2 in) 2 - 4 cm (1½ in) 2 - 2.5 cm (1 in) water-based craft paint: red, blue, green, orange, brown paintbrush water in a container misc. objects to represent planets (such as fruits, vegetables, candies, baking mixes) for *Just For Fun* section	ladybug or other small insect or animal to observe magnifying glass Just For Fun section: a second living thing or an object to observe

Materials

Quantities Needed for All Experiments

Equipment	Foods	Foods (continued)
bucket container, glass, tall, clear (approx. size: .5 liter [2 cups]) containers for mixing mud (about 1.75 liter [7 cups] size), 3 cups or small bowls (several) Experiment 13 suggestions: dominoes, blocks, electric car, electric train, tongue depressor, steel ball, other objects flashlight fork garden hose or funnel glass container, tall clear (approx .5 liter [2 cups]) Legos magnifying glass measuring cup measuring spoons pail (small) or plastic container pliers scissors shovel (small) or garden trowel spoon spoon or other implement for mixing mud trowel, garden	apple (2) baking soda (118 ml [8 Tbs]) or more banana (1-2) beans (several) cabbage, red (one head) cake mix and items needed to make the cake nuts, gumdrops, chocolate chips, and/or M&Ms to go in cake (see Experiment 16, Just For Fun) candy, color-coated such as M&Ms (several pieces) cheese (1 piece) crackers (1 or more) Experiment 9 Just For Fun suggestions: food items such as honey, berries, chopped fruit or vegetables, spices, herbs, cocoa, chocolate chips, etc. Experiment 21 Just For Fun: miscellaneous objects to use to represent planets (such as fruits, vegetables, candies, baking mixes, etc.) food items chosen by students	flour (60 ml [¼ cup]) food items chosen by students lemon or lime (1) lemon juice (180 ml [¾ cup]) marshmallows, large (1 pkg) marshmallows, small (1 pkg) milk (480 ml [2 cups]) orange (1) sugar, small quantity vinegar (180 ml [¾ cup]) or more water, tap yogurt, plain (.5 liter [2 cups])

Materials

Quantities Needed for All Experiments

Materials	Materials (continued)	Other
ball, ping-pong ball, rubber ball, tennis (2) balloon baseball or similar hard-centered ball basketball cardboard (stiff) or board (approx .3 meter x 1 meter [1 foot x 3 feet]) cotton balls (several) cups, clear plastic, 12 or more dirt, for mud pies (2 liters [8 cups]) or more dirt that contains rocks (.25 liter [1 cup]) feather (optional) glue or tape jar top, shallow marbles (2) misc. objects to be sorted into categories (see Exper. 7) paint, water-based craft: red blue green orange brown paints, watercolor paintbrush paper, several sheets paper to paint on, several sheets	pen, marking pencil pencils, colored pennies (optional) plastic bag, small playing cards (3) rocks (several) rubber bands string, 1 piece styrofoam balls (8): 1 - 10 cm (4 inch) ball 1 - 7.5 cm (3 inch) ball 2 - 5 cm (2 inch) balls 2 - 4 cm (1½ in) balls 2 - 2.5 cm (1 in) balls substances to test for acidity-basicity (see Exper. 5 Just For Fun) tape toilet paper tube, empty toothpicks, 1 box toy car water water in a container (to use with paints)	dark room internet access and/or reference books ladybug or other small insect or animal to observe living things to observe, such as an ant, frog, bird, cat, plant, or dog night sky, any night sky, clear non-living object to observe, such as a rock or piece of wood

Contents

INTRODUCTION

Experiment 1	Doing Science	1

CHEMISTRY

Experiment 2	Chemistry Every Day	6
Experiment 3	What Is It Made Of?	9
Experiment 4	Follow the Rules!	14
Experiment 5	What Will Happen?	21

BIOLOGY

Experiment 6	What Is Life?	26
Experiment 7	Where Does It Go?	29
Experiment 8	What Do You Need?	35
Experiment 9	Yummy Yogurt	40

PHYSICS

Experiment 10	Falling Objects	43
Experiment 11	Get To Work!	48
Experiment 12	Moving Energy in a Toy Car	53
Experiment 13	Playing With Physics	58

GEOLOGY

Experiment 14	Geology Every Day	63
Experiment 15	Mud Pies	66
Experiment 16	The Shape of Earth	70
Experiment 17	Mud Volcanoes	73

ASTRONOMY

Experiment 18	Observing the Stars	77
Experiment 19	Earth in Space	80
Experiment 20	Seeing the Moon	86
Experiment 21	Modeling the Planets	89

CONCLUSION

Experiment 22	Putting It All Together	92

CHEMISTRY

BIOLOGY

PHYSICS

GEOLOGY

ASTRONOMY

Experiment 1

Doing Science

Materials Needed

- colored pencils
- night sky
- a living thing to observe (frog, ant, plant, or other)

Objectives

This is an introductory experiment that gives students the opportunity to practice using a laboratory notebook for recording their experiments.

The objectives of this lesson are:

- To introduce students to the five different sections that appear in each experiment in the *Exploring the Building Blocks of Science Laboratory Notebook: Book 1.*
- To show students how to use the *Laboratory Notebook.*

Experiment

Introduction

Read this section of the *Laboratory Notebook* with your students and discuss any questions they may have.

I. Think About It

Note: In some experiments the *Think About It* section may come after the *Observe It* section.

Read this section of the *Laboratory Notebook* with your students.

Have the students think about what a night sky might look like. Encourage open inquiry with following questions:

- *What do you think the sky looks like at night?*
- *What color do you think it is?*
- *What do you think stars look like?*
- *What do you think the Moon looks like?*
- *Do you think the Moon stays the same all night or does it change?*
- *Do you think the sky stays the same every night or does it change?*

There are no "right" answers to these questions. Just allow the students to explore their own ideas about the night sky.

In the space provided in the *Laboratory Notebook,* help the students write or draw what they think they might see.

II. Observe It

Read this section of the *Laboratory Notebook* with your students.

Have the students observe a night sky. It can be any sky, with our without clouds, early or late in the evening.

Encourage open inquiry with following questions.

- *What does the night sky look like?*
- *What color is it?*
- *Do clouds change what you can observe?*
- *Can you see the Moon? What shape is it? What color? Can you see any patterns on the Moon?*
- *Can you see any stars? Are they in groups? Are they all the same size? The same color?*

In the space provided, help the students write or draw what they actually observe.

III. What Did You Discover?

Read this section of the *Laboratory Notebook* with your students.

❶-❻ Have the students answer the questions. These can be answered orally or in writing. Again, there are no right answers, and their answers will depend on what they actually observed.

IV. Why?

Read this section of the *Laboratory Notebook* with your students and discuss any questions that might come up.

V. Just For Fun

Read this section of the *Laboratory Notebook* with your students.

Help your students create their own experiment. In this section they get to be the teacher and tell you how to do an experiment using the five sections in the *Laboratory Notebook.*

Following are some suggestions for the experiment, or students can be encouraged to come up with their own idea. A living thing is suggested but not required.

- *Observe an ant.*
- *Watch a frog.*
- *Observe a plant and note its features.*

A template is provided in the *Laboratory Notebook* where students can write down the title of their experiment and then fill in instructions and questions for the various sections. Or they can orally instruct you how to walk through each of the five sections.

There are no right or wrong answers in this experiment. Encourage the students to think, to observe, and to use their imagination.

I. Think About It

Have the students come up with some instructions about what things the experimenter should think about before beginning the experiment. What might be observed?

Help them record their instructions for this section.

As the person performing the experiment, record your thoughts about what you might see.

II. Observe It

Have the students come up with some instructions about what is to be observed in their experiment.

Help them record their instructions.

As the person performing the experiment, record what you observed as you did the experiment.

III. What Did You Discover?

❶-❹ Guide the students in coming up with some questions to ask about what was observed during the experiment.

Help them record their questions.

Record your answers to their questions.

IV. Why?

In this section of the *Laboratory Notebook*, information is provided to the students about why certain things may have been observed during the experiment. Have the students think about why the things that were observed during this experiment may have happened. Then have the students write or draw their ideas in the space provided.

V. Just For Fun

Help the students come up with an extra activity that they think would be fun to have you try (or they might want to do it themselves instead). The activity should be related to the experiment they created. For example, if the experiment involves observing a plant, they might want to have you look for some plants that have leaves or flowers that are different from the plant in the experiment and then record the differences you observe. If they are having you observe an ant, they might want to have you imagine what the part of an anthill that is below the ground looks like and then draw it. If they are having you observe a frog, they might want to have you make up and draw an imaginary insect that you think would look tasty to a frog.

See what activity they can come up with and have them write or draw instructions for it. On the next page of the *Laboratory Notebook*, you can draw or write about the activity, or the students can do the activity themselves.

Experiment 2

Chemistry Every Day

Materials Needed

- watercolor paints
- water in a container
- paintbrush
- several pieces of paper to paint on
- scissors
- tape

Objectives

In this experiment students explore how chemistry is involved in activities they perform daily.

The objectives of this lesson are for the students to:

- Observe their activities.
- Make the connection that the activities they perform involve some aspect of chemistry.

Experiment

I. Think About It

Read this section of the *Laboratory Notebook* with your students.

❶-❷ Have the students think about and make a list of the activities they perform in a day. Some suggestions are:

- *Brushing their teeth.*
- *Shampooing their hair.*
- *Eating a cooked egg.*
- *Washing something with soap.*
- *Using a car or other motorized vehicle for transportation.*

❸ Guide the students in their exploration of whether or not these activities involve chemistry.

There are no right answers for these questions. Just allow the students to explore their own ideas.

II. Observe It

Read this section of the *Laboratory Notebook* with your students.

Have your students observe and make a list of everything they do during the course of one day. Have them be as specific as possible.

Have them observe when they are using products such as toothpaste, shampoo, soap, other cleaning products, or medications.

Have them observe the use of any tools or machines. Are they riding a bike, riding in a car, or using an electric powered tool?

III. What Did You Discover?

Read the questions with your students.

❶-❹ Have the students answer the questions. These can be answered orally or in writing. Again, there are no right answers, and their answers will depend on what they actually observed.

IV. Why?

Read this section of the *Laboratory Notebook* with your students.

Discuss any questions that might come up.

V. Just For Fun

Read this section of the *Laboratory Notebook* with your students.

❶-❹ Help your students use watercolors to observe how colors mix. Have them notice how colors change and which colors mixed together will make black.

❺ If the students are interested, they may enjoy experimenting with mixing different colors of their own choice to see what happens.

❻ Have the students cut out some examples of their paint mixtures and tape them in their *Laboratory Notebook* in the space provided. A hair dryer can be used to speed drying time.

Experiment 3

What Is It Made Of?

Materials Needed

- magnifying glass
- household items such as:
  cotton balls
  rubber bands
  pencil
- several food items such as:
  crackers
  cheese
  marshmallow
  beans
  color-coated candy
  (such as M&Ms)

Objectives

In this experiment students will learn how to make good observations.

The objectives of this lesson are:

- To have younger students make careful observations by noticing details.
- To help students develop a vocabulary to describe their observations.

Experiment

I. Think About It

Read this section of the *Laboratory Notebook* with your students.

Here the students will think about and describe the features of objects such as a cracker, a piece of cheese, a piece of candy. **Without allowing the students to look at the object**, name the object and have the students describe it, using both words and pictures. (They will observe the actual object in the following section.)

Direct their inquiry with questions. For example:

- *What color is a cracker?*
- *Is a cracker hard (like a plastic toy) or soft (like a feather)?*
- *Is a cracker large (like an elephant) or small (like a mouse)?*
- *Is a cracker smooth (like a marble) or rough (like sandpaper)?*
- *If you break a cracker, does it look the same on the inside as the outside?*

Using a cracker as an example, the students' answers may look something like the following example.

(Answers may vary.)

Write down the name of an object. Using words and drawings, describe any features you think it has.

cracker

brown	*round*	*scratchy*	*crumbly*

II. Observe It

Read this section of the *Laboratory Notebook* with your students.

Have the students look carefully at the object you've provided for them. Using the magnifying glass, have them examine the object and make careful observations about it. Ask them if the object looks different from what they thought it would.

Direct their investigation with questions such as the following, again using the cracker as an example.

- *Is the cracker as large (or as small) as you thought it would be?*
- *Is the cracker smooth or rough?*
- *What color is the cracker? Is it exactly brown (or white)? Does it have other colors in it?*
- *What happens to the cracker if you break it in half? Is it the same on the inside as on the outside?*
- *What does the cracker look like under the magnifying glass? Can you describe what you see?*

Often students discover that they have not seen or thought about some detail of a familiar object. For example, sometimes crackers have holes on the top. This may be a detail they have never noticed. Or there may be stripes or speckles in the cracker that they haven't observed before. Also, they can observe that some objects are the same on the inside as they are on the outside, like cheese and cotton balls, but other things are not the same on the inside and the outside, like color-coated candy or beans.

For this part of the experiment (using the cracker as an example), the students' answers may look something like this:

(Answers may vary.)

Write down the name of the object you thought about. Describe what you actually see, using words and drawings.

cracker

white and brown speckles	*holes on top*	*rough inside*	*seeds*

Have the students repeat this exercise with two or three more objects, first describing the item in the *Think About It* section without looking at the object and then using the *Observe It* section to record what they actually see. They can observe as many items as they want to, describing the item first without looking at it, and then carefully observing the item with a magnifying glass. They may want to choose their own items to observe.

III. What Did You Discover?

Read the questions with your students.

❶-❻ The questions in this section of the *Laboratory Notebook* can be answered verbally or in writing, depending on the writing ability of the students. Help the students think about their observations as they answer these questions. Have the students compare their list of descriptions of each object before and after they looked at it, and help them notice where an observation was the same as what they expected to see and where an observation differed from their expectations.

IV. Why?

Read this section of the *Laboratory Notebook* with your students.

Have a discussion with the students about why their observations may have been different from what they thought they would see.

V. Just For Fun

Have the students choose a person to compare to themselves. They may want to choose someone who is in the room so they can look at that person as they do this part of the experiment.

Guide the students in making observations about what is similar and different between themselves and the other person. Encourage the students to notice details, and have them record their observations. The students' answers will be based on what they actually observe, and there are no right answers to this experiment.

Experiment 4

Follow the Rules!

Materials Needed

- Legos
- small marshmallows, 1 pkg
- large marshmallows, 1 pkg
- toothpicks

Objectives

In this experiment, the students will explore the way the building blocks of matter (atoms) fit together to make molecules.

The objectives of this lesson are to help students understand that:

- Matter is made of smaller units called atoms that combine to form molecules.
- Atoms follow specific rules when forming molecules.

Experiment

Before the students begin the experiment, go through the following introduction with them.

① Place Legos on a table and have the students look at them. Help the students observe that each Lego has a certain number of holes on one side and the same number of pegs on the other side.

② Help the students understand that each Lego has only a certain number of pegs and a certain number of holes. Therefore, only a certain number of structures can be built by attaching more Legos to a foundational Lego block's pegs and holes.

③ Direct the students' inquiry with questions such as the following:

- *How many Legos can you attach to the 2-pegged Lego?*
- *How many Legos can you attach to the 4-pegged Lego?*
- *How many Legos can you attach to the 16-pegged Lego?*

④ Compare the Legos to atoms. Help the students understand that atoms are like Legos in that they can only hook to other atoms in certain ways.

I. Think About It

Read this section of the *Laboratory Notebook* with your students.

❶-❺ Have the students answer the questions in this section. There are no right answers. The objective is to have the students begin to think about molecules having specific structures.

II. Observe It

Read this section of the *Laboratory Notebook* with your students.

Save all the marshmallow molecules the students make. They will be used later in the experiment. Keep the molecules made without rules separate from those made with rules.

❶ Have the students make "molecules" with the marshmallows and toothpicks. First, they will make as many different molecules as they can without using any rules. Have them make several and then draw one of them. An example is shown.

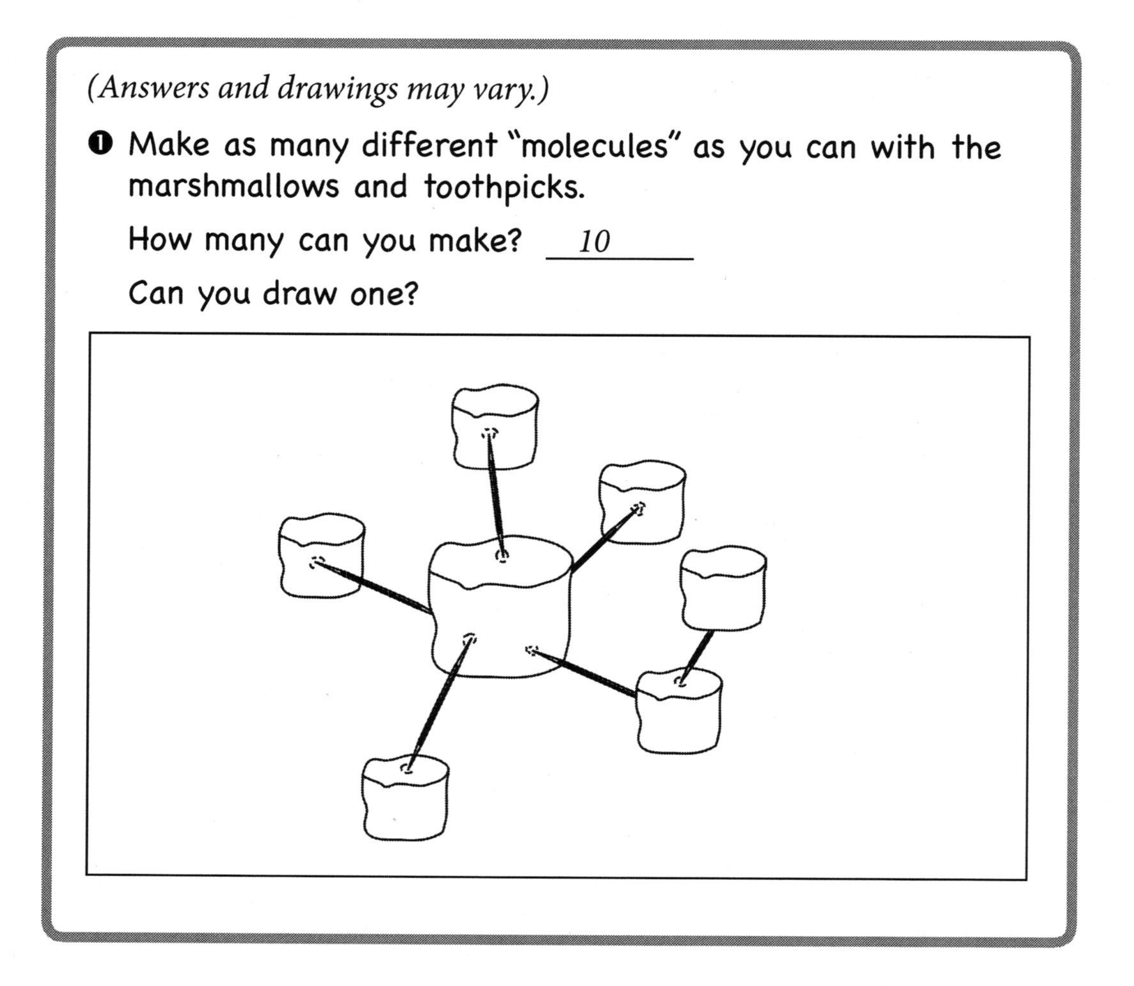

(Answers and drawings may vary.)

❶ Make as many different "molecules" as you can with the marshmallows and toothpicks.

How many can you make? *10*

Can you draw one?

❷ Next, have the students apply a rule to a big marshmallow. The rule is: ***Only three toothpicks can be put into a big marshmallow.*** They will put a small marshmallow on the other end of each toothpick. Have them make as many shapes as they can think of. Their answer may look something like this:

(Answers and drawings may vary.)

❷ This time you can put only three toothpicks into a big marshmallow. Following this rule, make as many different molecules as you can. A small marshmallow will go on the other end of each toothpick.

How many can you make? *5*

Can you draw one?

❸ Have the students apply a new rule to a big marshmallow. The new rule is: ***Only two toothpicks can be put into a big marshmallow.*** They will put a small marshmallow on the other end of each toothpick. Their answer may look something like this:

(Answers and drawings may vary.)

❸ Now you can put only two toothpicks into a big marshmallow. Following this rule, make as many different molecules as you can. A small marshmallow will go on the other end of each toothpick.

How many can you make? *3*

Can you draw one?

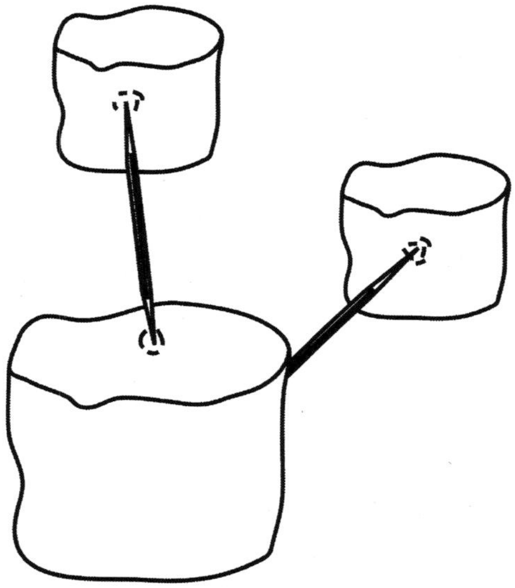

❹ Have the students apply the next rule to a big marshmallow. The rule now is: ***Only one toothpick can be put into a big marshmallow.*** They are to put a small marshmallow on the other end of the toothpick. Their answer may look something like this:

(Answers and drawings may vary.)

❹ Now you can put only one toothpick into a big marshmallow. Following this rule, make as many different molecules as you can. A small marshmallow will go on the other end of the toothpick.

How many can you make? *1*

Can you draw one?

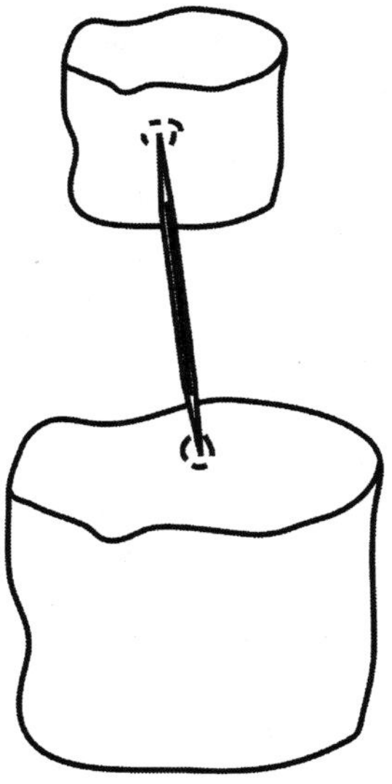

❺-❻ Have the students compare the marshmallow molecules they made with and without rules. Help them find two shapes that are the same as each other and two that are different.

Help them record their observations.

III. What Did You Discover?

Read the questions with your students.

❶-❹ Have the students answer the questions in this section. Help them see that by following rules they were able to make fewer molecules than when they didn't follow rules. Explain to them that it is important for atoms to follow rules when making molecules in order for the same substances to be formed consistently from the same combinations of atoms.

IV. Why?

Read this section of the *Laboratory Notebook* with your students.

Discuss the concepts presented in this section of the *Laboratory Notebook*. Help the students understand that when people don't follow rules, there is no order — there is chaos. Help them see that the same thing would be true if atoms did not follow rules when forming molecules.

V. Just For Fun

In this section, students are given the choice of thinking about an existing game they play or making up a new one.

If students choose to think about an existing game, they are asked to think about rules that they follow when they play this game. Encourage them to think about why the rules are important to the game. Help them record whatever rules they think of.

Instead of thinking about an existing game, students can make up their own game and its rules. Allow the students to be creative and use their imagination. Help them record the rules of their new game.

Encouraging students to think about rules and their effects is the focus of this exercise. There are no right answers to this section.

Experiment 5

What Will Happen?

Materials Needed

- 4 or more clear plastic cups or glasses
- marking pen
- measuring cup
- measuring spoons
- the following food items:
  lemon juice - 180 ml (3/4 cup)
  vinegar - 180 ml (3/4 cup)
  milk - 180 ml (3/4 cup)
  baking soda - 90 ml (6 Tbsp.)
  water - 180 ml (3/4 cup)

Just For Fun section

- baking soda
- vinegar
- sugar

Or

- 2 or more food items chosen by student

Objectives

In this unit students will observe chemical reactions.

The objectives of this lesson are:

- To have students observe changes that occur in some substances when they undergo chemical reactions.
- To help students understand that not all substances will react chemically when mixed together.

Experiment

Setup—to do ahead of time:

1. Label the plastic cups **A**, **B**, **C**, and **D**.
2. Pour 60 ml (1/4 cup) of lemon juice into **Cup A**.
3. Pour 60 ml (1/4 cup) of vinegar into **Cup B**.
4. Pour 60 ml (1/4 cup) of milk into **Cup C**.
5. Pour 60 ml (1/4 cup) of water into **Cup D** and add 30 ml (two tablespoons) of baking soda. Mix until the baking soda is completely dissolved.

(Each cup will be refilled twice with the same amount of the same liquid, or you can use a new cup each time, labeling it.)

Set the cups on a table. ***Do not tell the students what is in each cup.***

I. Think About It

Have the students examine the contents of each cup. They should observe the smell and color of the liquids and whether they are thick or thin. Have them write down what they observe, or record their answers for them.

II. Observe It

1. **Cups A and B**

   Ask the students what they think will happen if they add the contents of **Cup A** to **Cup B**. Have them write down their guess.

   Next, have them pour the contents of **Cup A** into **Cup B**. Have them observe and record what happens.

   Vinegar and lemon juice do not react, so they should not observe much happening.

Rinse the cups with water.

Refill **Cup A** with 60 ml (1/4 cup) of lemon juice and **Cup B** with 60 ml (1/4 cup)of vinegar.

❷ **Cups A and C**

Ask the students what they think will happen if they add the contents of **Cup A** to **Cup C.** Have them write down their answers.

Now have them pour the contents of **Cup A** into **Cup C.** Help them record what happens. The students will observe clumps forming in the milk as the lemon juice curdles the milk. This is a chemical reaction. The clumps are proteins in the milk that have been denatured (had their original properties changed) by the lemon juice. The clumps form a precipitate.

Rinse the cups with water.

Refill **Cup A** with 60 ml (1/4 cup) of lemon juice and **Cup C** with 60 ml (1/4 cup) of milk.

❸ **Cups A and D**

Ask the students what they think will happen if they add the contents of **Cup A** to **Cup D.** Have them write down their answers.

Have them pour the contents of **Cup A** into **Cup D.** Help them record what happens. They should observe a chemical reaction occurring between the lemon juice and the baking soda. The chemical reaction gives off bubbles which should be visible. If this does not happen, pour out the contents of **Cup D.** Make a new mixture of water and baking soda, and add twice as much baking soda. It doesn't matter whether or not all of the baking soda is dissolved. Then repeat the experiment.

Rinse the cups with water.

Refill **Cup D** with 60 ml (1/4 cup) of baking soda water.

❹ **Cups B and C**

Next, ask the students what they think will happen if they add the contents of **Cup B** to **Cup C.** Have them write down their answers.

Have them pour the contents of **Cup B** into **Cup C.** Help them record what happens. They should observe a chemical reaction similar to that of Step ❷. The vinegar should cause the milk to curdle.

Rinse the cups with water.

Refill **Cup B** with 60 ml (1/4 cup) of vinegar and **Cup C** with 60 ml (1/4 cup) of milk.

❺ **Cups B and D**

Next, ask the students what they think will happen if they add the contents of **Cup B** to **Cup D**. Have them write down their answers.

Have them pour the contents of **Cup B** into **Cup D**. Help them record what happens. They should observe a chemical reaction similar to that of Step ❸. The vinegar and baking soda should react, and the mixture should give off bubbles.

Rinse the cups with water.

Refill **Cup D** with 60 ml (1/4 cup) of baking soda water.

❻ **Cups C and D**

Ask the students what they think will happen if they add the contents of **Cup C** to **Cup D**. Have them write down their answers.

Have them pour the contents of **Cup C** into **Cup D**. Help them record what happens. They should not observe any chemical reaction taking place. The baking soda will not make the milk curdle, nor will there be any visible bubbles or any other signs of a chemical reaction taking place.

Have the students pour out the contents of **Cup D** and clean up the experiment space.

Summary

Have the students summarize their results. The answers to questions ❶-❻ are provided below.

❶ Did lemon juice (**A**) react with vinegar (**B**)? *no*

❷ Did lemon juice (**A**) react with milk (**C**)? *yes*

❸ Did lemon juice (**A**) react with baking soda (**D**)? *yes*

❹ Did vinegar (**B**) react with milk (**C**)? *yes*

❺ Did vinegar (**B**) react with baking soda (**D**)? *yes*

❻ Did milk (**C**) react with baking soda (**D**)? *no*

III. What Did You Discover?

Help the students answer the questions in this section of the *Laboratory Notebook*. Their answers may vary depending on what they observed.

IV. Why?

Read this section of the *Laboratory Notebook* with your students.

Discuss the similarities and differences between the four liquids. Explain that vinegar and lemon juice are made of similar kinds of molecules, but milk and baking soda are not similar to vinegar and lemon juice. Although lemon juice and vinegar are not identical, they behave in similar ways because they are composed of similar kinds of molecules.

V. Just For Fun

Students can try mixing baking soda, sugar, and vinegar together in a cup. Let them experiment with how much of each substance to use. Adding sugar should result in a more impressive reaction. This experiment may be repeated using different amounts of the various substances.

Alternatively, students can look around the kitchen to find two other foods to try mixing together to see if they will react. This experiment may be repeated with other food combinations.

Have the students record their results in the space provided.

Experiment 6

What Is Life?

Materials Needed

- non-living object to observe, such as a rock or piece of wood
- living thing to observe, such as an ant, frog, bird, cat, or dog
- colored pencils

Objectives

In this experiment students will explore the differences between living and non-living things.

The objectives of this lesson are for students to:

- Observe features of living and non-living things.
- Think about the definition of life.

Experiment

I. Think About It

Read this section of the *Laboratory Notebook* with your students.

❶-❹ Help the students think about and answer the questions in this section. There are no right answers for these questions. Just allow the students to explore their own ideas about life and non-life.

II. Observe It

Read this section of the *Laboratory Notebook* with your students.

Help your students find a living thing and a non-living thing to observe. Have them write the name of each item in the space provided. In one column they will list or draw their observations for the living thing and in the other column their observations for the non-living thing.

Help them explore each item by asking the following questions.

- *Can the item move?*
- *Does the item breathe?*
- *Does the item consume food?*
- *Can the item reproduce itself?*

In the space provided have the students write or draw what they see.

III. What Did You Discover?

Read the questions with your students.

❶-❸ Have the students answer the questions. These can be answered orally or in writing. There are no right answers, and their answers will depend on what they actually observed.

IV. Why?

Read this section of the *Laboratory Notebook* with your students.

Discuss any questions that might come up.

V. Just For Fun

Read this section of the *Laboratory Notebook* with your students.

Have the students imagine what life on another planet might look like. Help them think about what characteristics they might include, such as how the creatures would move, what sort of appendages they would have and how many, their skin or fur color, the environment they'd live in, etc.

Have them write about and/or draw their imaginary living things. They can assign a name to the planet and name their creatures.

Experiment 7

Where Does It Go?

Materials Needed

- cotton balls
- rubber ball
- tennis ball
- banana
- apple
- rocks
- Legos
- other objects
- colored pencils

Objectives

In this experiment the students will try to sort objects into different groups according to their characteristics.

The objectives of this lesson are:

- To help students understand that there are different ways to sort objects.
- To have students develop a vocabulary to describe the objects they observe.

Experiment

I. Observe It

In this section the students will make careful observations for each of the objects they have collected.

Read this section of the *Laboratory Notebook* with your students.

❶ Help the students collect objects to observe.

❷ Put the objects on a table and have the students look carefully at each item.

Help them observe different details such as size, color, shape, and texture for each item. Use questions to help them describe the object.

- *What color is a cotton ball?*
- *What color is a banana?*
- *What is the shape of a cotton ball?*
- *What is the shape of a rock?*
- *How would you describe the surface of a tennis ball?*

❸ Encourage students to use both words and pictures to describe each object. Have them use as many different describing words as possible for each item. Their answers may look something like this:

(Answers will vary.)

cotton ball

fuzzy *round* *soft* *white*

tennis ball

fuzzy *round* *hard* *yellow*

(Answers will vary.)

rock

hard	*gray*	*smooth*	*heavy*

building block

square	*hard*	*blue*	*plastic*

II. Think About It

Read this section of the *Laboratory Notebook* with your students.

1. Have the students look at the different objects and the different ways they have described the objects. Help them think about how they might group the objects according to their descriptions.

❷ Have the students think of five different groups and then write the names of the groups in the gray boxes. Next, have them sort the objects they have collected into the different groups. Each object can only go into **ONE** group.

Their answers may look something like this:

(Answers will vary.)

round	*yellow*	*small*	*hard*	*white*
tennis ball	*banana*	*rock*	*block*	*cotton ball*
	apple		*rubber ball*	

❸-❹ Help the students notice that some items can fit into more than one group. For example, if they chose the groups round and yellow, a tennis ball can fit into both groups. Have the students think about how they might rearrange the groups, picking different items that go into each group. They can re-sort their items into the groups they've already chosen, or they can pick new groups.

Help them notice features that are similar and different between the objects by asking questions such as:

- *Is a rubber ball larger or smaller than a cotton ball?*
- *Is a rubber ball harder or softer than a cotton ball?*
- *Is a rock like a banana? Why or why not?*
- *Is a tennis ball similar to a banana? Why or why not?*

There are no "right" answers, so encourage the students to think about all the different ways they may want to sort the items.

III. What Did You Discover?

Read the questions with your students.

❶-❹ The questions can be answered verbally or in writing, depending on the writing ability of the student. With these questions, help the students think about their observations. Again, there are no "right" answers to these questions, and it is important for the students to write or discuss what they actually observed.

IV. Why?

Read this section of the *Laboratory Notebook* with your students.

It is important for students to understand that science is often a dynamic endeavor, and the "answers" that science provides can sometimes change. The identification and grouping of living things can be fairly complicated, and determining exactly which group a living thing belongs to is not trivial. There are different criteria used to group living things. Most living things are first grouped according to the types of cells they have — plant cells, animal cells, bacterial cells, etc. Once the organism is grouped according to cell type, then the scientist looks for other features to use in categorizing the organism.

V. Just For Fun

Read this section of the *Laboratory Notebook* with your students.

Help the students think about how they would categorize the new creature according to the features and groups given. There are no right or wrong answers in this section.

Encourage students to use their imagination in creating this new creature.

Experiment 8

What Do You Need?

Materials Needed

- internet access and/or reference books
- colored pencils

Objectives

In this experiment, students will explore different jobs their parents do in the home. They will also explore the steps it takes to bring a single tool into the house so that a parent can do their job. This will help students understand how a cell has many different parts that rely on each other and work together to allow the cell to live.

The objectives of this lesson are to help students:

- Observe some of the many different jobs needed to run a household.
- Understand how all of the jobs their parents do are connected to other jobs performed by other people.
- Understand how different parts of a cell work together and perform different functions.

Experiment

I. Observe It

Read this section of the *Laboratory Notebook* with the students.

❶ Have your students follow their parent around and observe the jobs that are done. You might want to pick a day where your students can observe many different jobs, such as doing laundry, cooking a meal, washing the dishes, repairing a broken door, hanging a picture, cleaning the yard, or mowing the lawn.

Have the students bring their *Laboratory Notebook* with them as they observe their parent so they can record several jobs as they observe them being done. Their answers may look something like this:

(Answers will vary.)

Job	*cooking breakfast*
Job	*fixing the broken stove*
Job	*planting the garden*

❷ Have your students pick one of the jobs on their list. Have them think about and then write down or draw all of the items needed to do that job. Their answers may look something like this:

(Answers will vary.)

Job *cooking breakfast*	
Items Needed	
eggs	*bowl*
whisk	*frying pan*
butter	*salt*
pepper	*green chili (for NM residents)*

❸ Have your students draw a picture of the person doing the selected job. Have them include the tools that would be used to do the job.

❹ Have the students pick one of the tools from Step ❸ and draw it in detail in the space provided.

II. Think About it

Read this section of the *Laboratory Notebook* with your students.

❶-❻ Have the students answer the questions about the item they have selected. They may not be able to answer the questions exactly (they may not know exactly where the bowl or the whisk was purchased). Help them come up with general answers if needed ("grocery store" or "hardware store" is enough of a description).

Their answers may look something like this:

❶ What is the item?

whisk

❷ Where did the item come from?

grocery store

❸ How did the item get there?

on a truck

❹ Who made the item?

the whisk factory

❺ What is the item made of?

steel

❻ Where does the material that makes the item come from?

from iron ore

III. What Did You Discover?

Read this section of the *Laboratory Notebook* with your students.

❶-❻ Guide the students in answering the questions in this section. Help them think about all of the items people, including their parents, use to do different jobs. Have them think about all of the people it takes to make the items their parents use to do a job.

IV. Why?

Read this section with your students.

This experiment helps students have a better understanding of the many jobs people do and how many items are needed for them to be able to do those jobs. Help the students connect what they observed in this experiment with what goes on inside a cell. Help them see that a cell operates in a way that is similar to a city.

In order for someone to have a tool to do a job in a house, that tool needs to come from other people in other cities doing different jobs. Explain that there are "jobs" that proteins do inside cells so that the cells can live. Proteins inside the cells do these "jobs." In order for a protein to do its job, it depends on other proteins to do their jobs. For example, for a protein to move molecules from one place to another, other proteins are required to make the molecules that are to be moved.

Help the students see that inside a cell there is a very sophisticated network of proteins and other molecules that do the different jobs that make it possible for the cell to live.

V. Just For Fun

Read this section with your students.

Have the students select an item to observe. Guide their observations of the materials contained in the object they choose.

By looking up some of the ways materials, such as steel, are made, the students will gain a better understanding of how many people it takes to produce an item, such as a cooking whisk. Help the students search the internet or look at reference books to find the information.

Experiment 9

Yummy Yogurt

Materials Needed

- .25 liter (1 cup) regular milk
- .5 liter (2 cups) plain yogurt
- fork
- spoon
- several cups or small bowls
- food items such as honey, berries, chopped fruit or vegetables, spices, herbs, cocoa, chocolate chips, etc. (see *Just For Fun* section)

Objectives

In this experiment, students will examine the differences between yogurt and milk.

The objectives of this lesson are for students to:

- Make comparisons, observing differences and similarities.
- Examine how bacterial cultures change the consistency of milk.

Experiment

I. Think About It

Read this section of the *Laboratory Notebook* with your students.

Have the students think about bacteria and how bacteria live in many different places. ***Before having the students look at the yogurt and milk***, help them think about what the differences between yogurt and regular milk might be. Guide their inquiry with questions such as the following:

- *What color do you think milk is?*
- *What color do you think yogurt is?*
- *Do you think yogurt tastes different from milk? If so, what do you think the difference in taste would be?*
- *Do you think you can you eat yogurt with a fork? Why or why not?*
- *Do you think you can you eat milk with a fork? Why or why not.*

There are no right answers for these questions. Just allow the students to explore their own ideas about the differences between yogurt and milk.

II. Observe It

Read this section of the *Laboratory Notebook* with your students.

Provide about .25 liter (1 cup) of milk and .25 liter (1 cup) of yogurt to the students. Have them observe the color, smell, taste, and consistency of each. Have them use a fork, a spoon, and their fingers to test the consistency.

Help students record their observations in the columns provided in their *Laboratory Notebook.*

III. What Did You Discover?

Read the questions with your students.

❶-❹ Have the students answer the questions. These can be answered orally or in writing, depending on the writing ability of the student. Again, there are no right answers, and their answers will depend on what they actually observed.

IV. Why?

Read this section of the *Laboratory Notebook* with your students.

Discuss any questions that might come up.

V. Just For Fun

Read this section of the *Laboratory Notebook* with your students.

Have the students think about what foods they might add to yogurt to change its flavor, color, texture, and/or smell. Have them look around the kitchen to find some food items that look interesting to them. The objective is for students to try different mixtures and make observations about the outcomes. Some of their ideas may result in mixtures that you know won't taste good, but let them try these anyway.

Help the students mix their chosen food item into a small portion of yogurt and then taste it. Have them record their observations.

Guide the students' inquiry by asking questions about what they observe.

Experiment 10

Falling Objects

Materials Needed

- 2 tennis balls
- other objects such as:
  apple
  orange
  rubber ball
  cotton ball or feather

Objectives

In this experiment students will try to determine if Galileo was right.

The objectives of this lesson are to have students:

- Compare their own observations with a scientific discovery.
- Compare different observations.

Experiment

I. Observe It

In this section students will observe how two objects fall when they are released at the same time.

Read this section of the *Laboratory Notebook* with your students.

❶ Have the students hold a tennis ball in each hand with their arms outstretched at chest level.

❷ Have the students release the two tennis balls at the same time.

❸ Help them observe how the objects land on the ground. Guide their inquiry with the following questions.

- *Did both objects land at the same time?*
- *Is one object heavier or lighter than the other object?*
- *Do you think it matters how high you hold the objects? Why or why not?*
- *Do you think the shape of the object matters? Why or why not?*

❹ Help the students record their observations in the *Observe It* section of their *Laboratory Notebook* (see example on next page).

❺ Have the students repeat the experiment using different combinations of objects. Have them compare at least four different pairs of objects. For each set of objects, help the students record their observations in the boxes provided.

In the spaces next to **Object 1** and **Object 2**, have the students write the names of the objects they will be dropping. Then have them draw or write a description of what they see. Help them make good observations by asking questions such as:

- *How heavy does Object 1 feel in your hand?*
- *How heavy does Object 2 feel in your hand?*
- *Does Object 1 feel heavier or lighter in your hand than Object 2?*
- *Is it easy to release both objects at the same time? Why or why not?*
- *Describe the shape of Object 1.*
- *Describe the shape of Object 2.*

Object 1 *apple*

Object 2 *tennis ball*

(Answers will vary.)

The apple feels heavier in my hand than the tennis ball.

When I drop the objects, I have a hard time seeing which one lands first. They look like they land together, but I am not sure.

I can release the objects at the same time from my hands.

The apple is a different shape than the tennis ball, but they still land at the same time.

II. Think About It

Read the questions with your students.

❶ Have the students think about their experiment and make observations about how easy or difficult it was to determine if the objects fell at the same speed.

- *Could you see the objects fall to the floor?*
- *Could you determine if both objects hit the floor at the same time?*
- *Was it easy or difficult to release the objects from both hands at the same time?*

❷ Help the students think about ways to vary their experiment.

- *If they can't see the objects fall to the floor, maybe they can get a parent, sibling, or friend to make the observations.*
- *Could they use a mirror to more easily see the objects hit the floor?*
- *What would happen if the objects were dropped from a greater height? Students could stand on a sturdy chair or bench to see if greater height makes a difference.*
- *What would happen if the objects were dropped from a lower height? Students could release the objects at waist or knee height.*

❸ Help the students repeat the experiment for one of their object pairs. Have them vary only one parameter at a time for one set of objects. For example, they may want to hold the objects higher and have a friend observe how they fall. But this is changing two parameters (the height and the observer), and if the results varied, the students couldn't tell which change to the experiment brought about the variation. Explain to them that scientists try to change only one parameter at a time so that they can make comparisons to previous experiments, noting what difference, if any, changing that one parameter had on the experimental results.

❹ Help the students record their observations.

III. What Did You Discover?

Read the questions with your students.

❶-❹ The questions can be answered verbally or in writing. With these questions help the students think about their observations. There are no right answers to these questions, and it is important for the students to write or discuss what they actually observed.

IV. Why?

Read this section of the *Laboratory Notebook* with your students.

Help the students understand that two objects of different weights will fall to Earth at the same speed. Both objects have the same amount of gravity pulling on them at the same time, so both objects start with the same force. Because both objects have the same force, they will both fall to the Earth at the same speed since the speed of an object is independent of its weight.

V. Just For Fun

Read this section of the *Laboratory Notebook* with your students.

The students can try an experiment with an object that is much lighter than the objects they have been using. A cotton ball or a feather would work. They will discover that if an object is too light, it will float to the ground and not fall at the same time as a heavier object. However, tell them that if the two objects are put in a vacuum, they will fall at the same time. Even an apple and a feather or cotton ball will fall at the same time. In a vacuum there is no air resistance. Outside the vacuum, the air pushes up on the cotton ball, and because the cotton ball is light enough, the air will slow it down.

Experiment 11

Get To Work!

Materials Needed

- large marshmallow
- tennis ball
- objects such as:
  rubber ball
  lemon or lime
  rock
  banana
- pliers

Objectives

In this experiment students will explore the concept that work happens when energy is used to create a force that moves an object, changes its shape, or changes its velocity.

The objectives of this lesson are for students to:

- Explore the concepts of work, force, and energy.
- Make careful observations.

Experiment

I. Observe It

In this part of the experiment students will apply a force to different objects and compare the results.

Read this section of the *Laboratory Notebook* with your students.

❶ Have the students observe a marshmallow. Have them write or draw a description of the size, shape, and color of the marshmallow in the "Before" section section of their *Laboratory Notebook* (see example on following page).

❷-❹ Have the students hold the marshmallow in one hand and then press the marshmallow with the palm and fingers of their hand. Use the following questions to help the students observe how much effort they used while squeezing the marshmallow.

- *Was it easy or difficult to squeeze the marshmallow?*
- *Were you able to squeeze the marshmallow completely? Why or why not?*
- *Would you describe the marshmallow as hard or soft?*

❺ Help the students record their observations in the "After" section of their *Laboratory Notebook* (see example on following page).

❻ Have the students repeat this exercise with several other objects.

Example

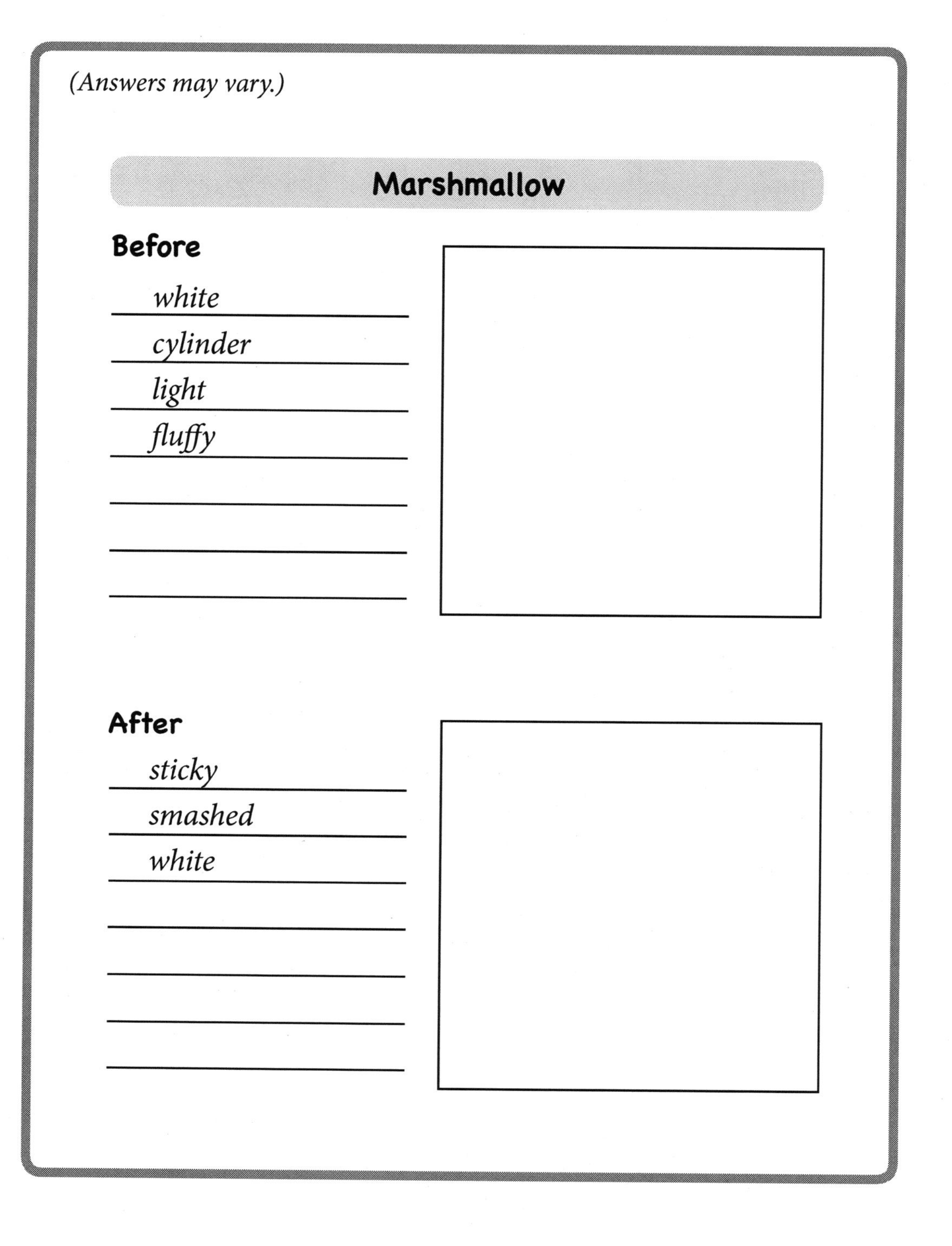
(Answers may vary.)
Marshmallow
Before
white
cylinder
light
fluffy
After
sticky
smashed
white

II. Think About It

Read the questions with your students.

❶ Using questions like the following, help the students think about their experiment and any observations they might have made about the objects they squeezed.

- *How did the objects feel in your hands? Were some objects heavy? Some light?*
- *Have the students discuss the size, shape, and color of the different objects.*
- *Were some objects difficult to hold? Did all the objects fit in your hand? Were some too large to hold?*

❷ Have the students discuss how difficult or easy it was to squeeze the objects.

❸ Help the students create a summary of their observations. There are two columns in the *Laboratory Notebook*— one labeled "Easy to Squeeze" and the other labeled "Hard to Squeeze." These are very general categories but will give the students a way to sort the different objects.

❹-❺ Review the chapter in the *Student Textbook* with the students and go over the summary statements. Have the students recall that work is what happens when a force moves an object, changes its shape, or changes how fast or slowly an object is moving.

Help them make the following connections.

1. *Their hand uses energy (from their body) in order to squeeze the object.*
2. *When squeezing the object, their hand generates a force on the object.*
3. *The object either changes shape as the force is applied or does not change shape.*
4. *The amount of work done is related to how much the object changes shape. The more the object changes shape, the more work is done.*

Help the students determine when they did the most work by squeezing the objects with their hands. Without telling them the answer, again remind them that the amount of work that is done depends on the amount the object changes shape. This means that even if they squeeze an object with all their effort, using lots of energy and applying lots of force, if the object does not change shape, they do no work.

Have the students circle the object they believe had the most work done to it when it was squeezed.

Have the students put a rectangle around the object they believe had the least work done to it when it was squeezed.

III. What Did You Discover?

Read the questions with your students.

❶-❺ The questions can be answered verbally or in writing. With these questions, help the students think about their observations. There are no right answers to these questions, and it is important for the students to write down or discuss what they actually observed. Help them explore how the answer they got may be different from what they thought might happen. If they guessed that more force would equal more work, they may be surprised to discover that this is not necessarily true.

IV. Why?

Read this section of the *Laboratory Notebook* with your students.

Discuss with the students how their body can be used as a tool to measure force and energy. Because their hands have nerve endings, they can sense the hardness or softness of different objects. Also, they can observe their muscles and breath as they squeeze different objects. This gives them an idea of how much energy is needed to generate different forces. The harder they squeeze an object, the more force they are generating.

Discuss with the students that the use of more force and more energy does not necessarily mean they did more work. The shape of an object has to change before work is done. If a force is applied, say to a rock, but the shape of the object does not change, no work is done. Discuss how it may feel as if they did work, but without an observable outcome, no work was done.

V. Just For Fun

Read this section of the *Laboratory Notebook* with your students.

Without a pair of pliers, it would be difficult for most young students to squeeze a tennis ball. However, with a pair of pliers it becomes much easier. The pliers are a form of tool called a lever, and they actually multiply the amount of force exerted on the tennis ball. That is, with a little applied force on one end (the handle) there is much greater force applied at the other end (the tennis ball).

Experiment 12

Moving Energy in a Toy Car

Materials Needed

- toy car
- stiff cardboard or board (approximately .3 meter wide x 1 meter long [1 foot x 3 feet])
- several marshmallows

Optional
(see *Just For Fun* section)

- pennies
- tape

Objectives

In this experiment students will explore how one form of energy can be converted to another form of energy. They will observe the conversion of gravitational stored energy into kinetic energy.

The objectives of this lesson are to have students:

- Observe how one form of energy is converted to another form of energy.
- Learn how to collect data and create a table of their results.

Experiment

I. Observe It

Students will perform a simple experiment to explore the conversion of gravitational stored energy to kinetic energy. By using a toy car, they will observe how increasing the gravitational stored energy in the car will give it more kinetic energy (moving energy).

Read this section of the *Laboratory Notebook* with your students.

❶ Have the students place the board or cardboard sheet on the ground with the toy car at one end.

❷ Even though the car is not moving, have them write or draw their observations. You can explain to the students that this is the "starting point" in the experiment. Even though nothing is happening, scientists always record their observations at a starting, or reference, point. A starting, or reference point, gives scientists an observation with which to compare observations that follow.

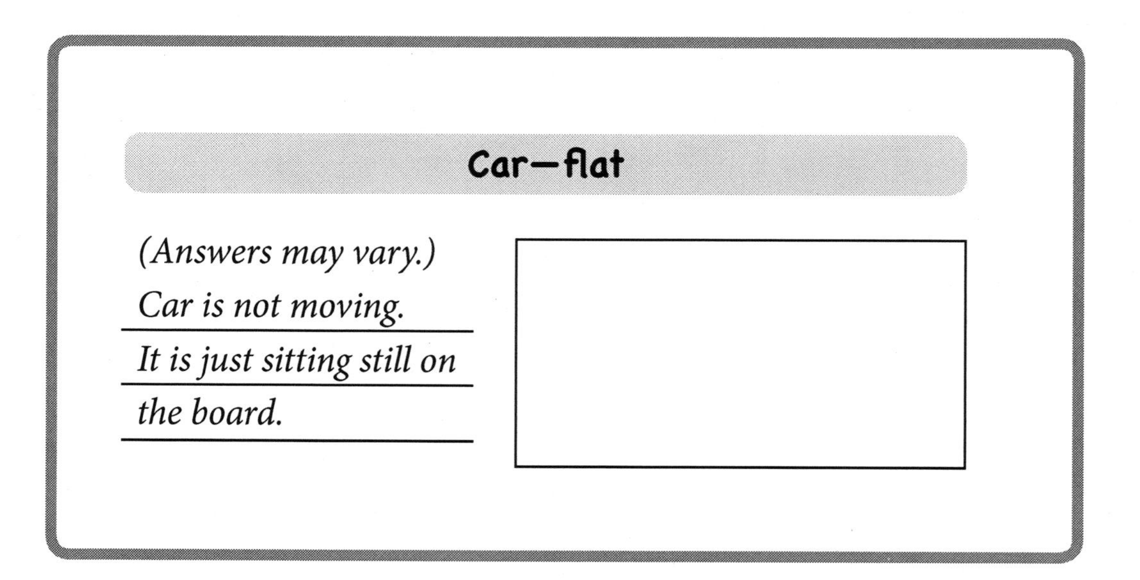

❸ Have the students stand at the end of the cardboard sheet or board on which the toy car is placed. Make sure the toy car is facing toward the other end of the cardboard sheet or board. Have the students lift the cardboard sheet or board to the level of their ankles, making a ramp. The toy car may or may not move at this height.

❹ Have the students record their observations. Guide their observations with the following questions:

- *When it starts out, is the toy car higher or lower now that you have lifted the board to your ankles?*
- *Did the toy car move? If so how far?*
- *If the toy car moved, how fast or slowly did it move?*
- *If the toy car moved, did it make it all the way to other end of the ramp?*

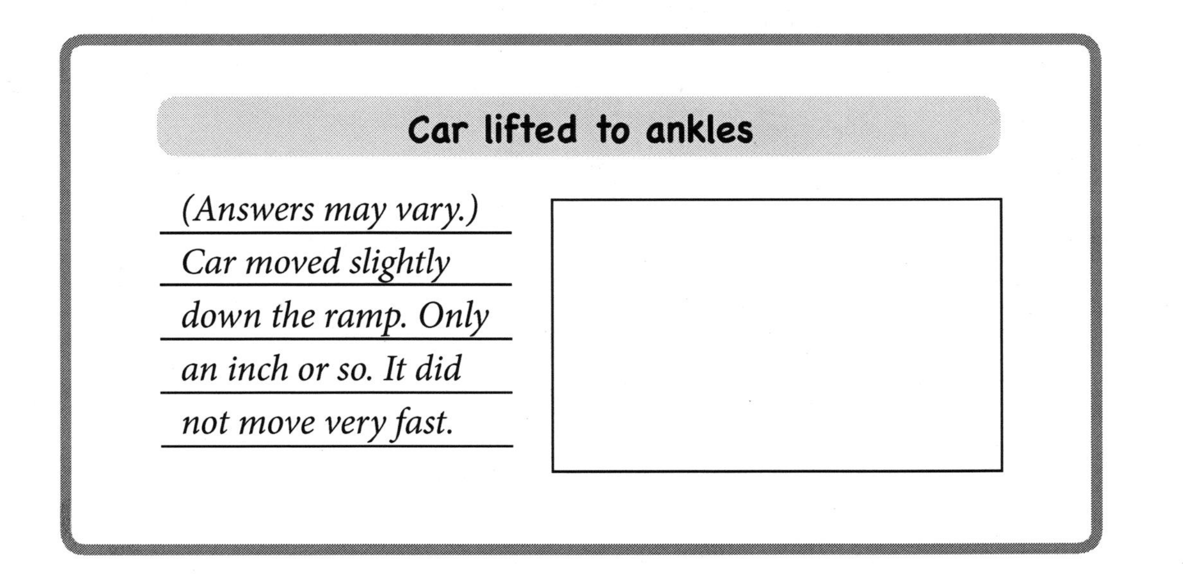

❺-❿ Repeat Steps ❸-❹ of the experiment for the car lifted to the knees, then the hips, and then the chest.

Collect Your Results

Have the students summarize their results in the table provided.

II. Think About It

Read this section of the *Laboratory Notebook* with your students.

❶ Help the students think about their experiment and any observations they made about the toy car and its movement. The following questions may be used to guide discussion:

- *Did the toy car move easily once the board was lifted?*
- *Were the wheels sticky, or did they turn easily?*
- *Do you think the car was heavy enough to move down the board?*
- *Was the car made of plastic or metal?*
- *Do you think what the car was made of would make a difference in the results?*
- *Was the board or cardboard sheet smooth or rough?*
- *Do you think what the board was made of made a difference in your experiment?*

❷ Review with the students the part of the section of the *Student Textbook* that covers gravitational stored energy. Any object that is elevated from the ground has some amount of gravitational stored energy. When the object falls to the ground, the gravitational stored energy is converted to kinetic energy. Objects that are higher than other objects have more gravitational stored energy.

❸ Have the students guess which car height had the least amount of gravitational stored energy and which height had the greatest amount of gravitational stored energy.

❹ The car had the least amount of gravitational stored energy when it was sitting flat on the ground. It had zero gravitational stored energy because it was at ground level. Have the students put a circle around this item.

❺ The car had the greatest amount of gravitational stored energy when it was lifted to the chest. Have the students draw a rectangle around this item.

Car flat

Car lifted to ankles

Car lifted to knees

Car lifted to hips

Car lifted to chest

III. What Did You Discover?

Read the questions with your students.

❶-❺ The questions can be answered verbally or in writing. With these questions help the students think about their observations. There are no "right" answers to these questions, and it is important for the students to write or discuss what they actually observed.

Help the students explore how the answers they got may be different from what they thought might happen. Help them understand the concept of converting gravitational stored energy to kinetic energy. The more gravitational stored energy the car has, the more kinetic energy it will generate as it moves down the ramp.

IV. Why?

Read this section of the *Laboratory Notebook* with your students.

Discuss with the students how the toy car was able to move once it was lifted from the ground. Help the students understand that as their body lifted the ramp, the car gained gravitational stored energy by being moved to a greater height. Help the students see that as they lifted the toy car higher, they added more gravitational stored energy to the car.

V. Just For Fun

Read this section of the *Laboratory Notebook* with your students.

Have the students observe how kinetic energy can do work. As they raise the ramp and the toy car moves, the toy car's gravitational stored energy is converted to kinetic energy. Work is done as the car moves and as the marshmallow is smashed.

Help the students qualitatively determine how much kinetic energy is needed to smash a marshmallow. They may have to lift the car and ramp higher than their heads. Also, it is possible that no matter how high they lift the ramp, the marshmallow won't smash. If this happens, you can ask them what they think might happen if they used a heavier car or taped some pennies to the car they are using to make it heavier.

Have the students test one change to the experiment at a time so they can tell which change brought about a different result.

Have them record their results in the box provided.

Experiment 13

Playing With Physics

Materials Needed

- 2 marbles
- 3 playing cards
- shallow jar top
- vinegar
- baking soda
- measuring spoons

Suggested

- dominoes
- blocks
- electric car
- electric train
- marshmallow
- tongue depressor
- steel ball
- other objects chosen by the student

Objectives

In this experiment students will set up several series of experimental events to explore how energy is converted from one form to another.

The objectives of this lesson are to have students:

- Combine several experiments into a series.
- Observe how energy is converted from one form to another form.

Experiment

I. Observe It

In this section students will use two marbles to design a simple experiment to explore kinetic energy.

❶ Have the students place one marble on the floor. Have them sit some distance away from the stationary marble and roll a second marble into it.

❷ Help the students draw what happens to both marbles. If the marbles are of equal size, the first marble will likely stop, and the second marble will begin to roll when the first marble hits it. The kinetic energy of the first marble is converted into kinetic energy in the second marble.

❸ Have the students arrange three playing cards to make a small card house. The students may need some help stacking the cards.

❹ Have the students roll a marble, hitting the card house. Have them draw what happens to both the card house and the marble. The cards should fall down, and the marble will likely stop.

❺ Have the students fill a shallow jar top (such as one from a pickle jar) with vinegar.

❻-❼ Have the students add 15 ml (1 tablespoon) of baking soda to the vinegar and have them draw what happens. They should observe bubbles being released from the chemical reaction.

❽-❾ Have the students rinse out the jar top and refill it with vinegar. Then have them make a card house above the jar top and place 15 ml (1 tablespoon) of baking soda on top of the card house. Next, they will tip the card house with their fingers until the baking soda falls into the vinegar. Have them record their observations.

⑩ Now have the students assemble all of these steps into a short series. Their setup should look something like this:

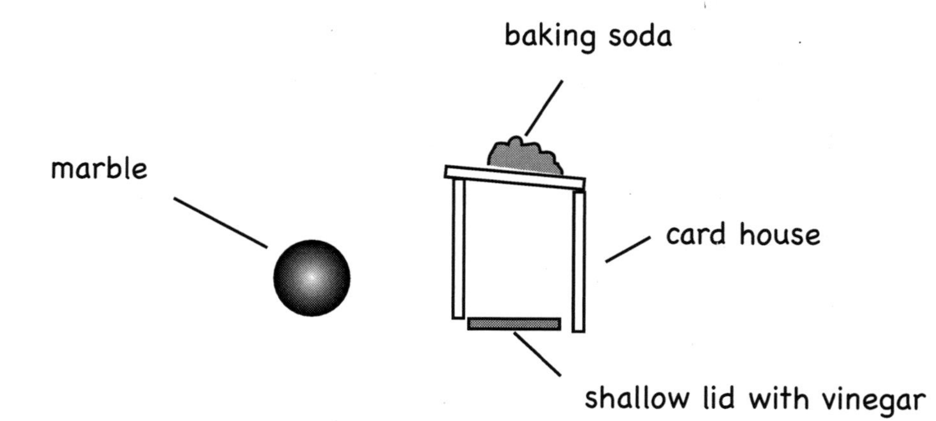

Have the students roll a second marble into the marble that is sitting close to the card house. This should cause the marble near the card house to roll into it and knock it down, causing the baking soda to fall into the shallow lid that contains the vinegar, resulting in a chemical reaction.

Have the students record their observations.

II. Think About It

❶ Help the students think about the different types of energy they used in this short series of smaller experiments. You can help them think about energy with the following questions.

- *Kinetic energy is the energy of something moving. What was moving in the experiment with two marbles?*
- *What was moving in the second experiment with a card house and a marble?*
- *Did anything move when you added vinegar to baking soda?*
- *What type of energy do you think is used when vinegar is combined with baking soda?*

2. Review Chapter 13, Section 13.1, *Energy to Energy* in the *Student Textbook*. Have the students think about how energy is converted from one form to another as shown in the illustration given in the book. Explain how energy is never created or destroyed, just converted from one form to another.

3. Have the students list the type of energy they think the object started with and the type of energy they think it was converted into. Have them fill out the chart using the energy labels given.

Object	Started With	Converted To
marble-->marble	*kinetic energy (rolling) marble*	*kinetic energy (rolling) marble*
marble-->card house	*kinetic energy (rolling)*	*kinetic energy (falling)*
card house upright--> card house falling	*stored energy (gravitational)*	*kinetic energy (falling)*
baking soda or vinegar -->baking soda + vinegar	*stored energy (chemical)*	*chemical energy*

III. What Did You Discover?

The questions can be answered verbally or in writing. With these questions, help the students think about their observations. There are no "right" answers to these questions, and it is important for the students to write or discuss what they actually observed. Help them explore how the answers they got may be different from what they thought might happen.

Help the students answer the questions about what happened to the energy at each stage of the series of experiments. Have the students think about what might have happened to all the energy at the end of the experiment. Where did it go?

IV. Why?

In this experiment the students explored how energy is converted from one form to another. The students observed how the kinetic energy of one marble is converted into the kinetic energy of another marble. The students also observed how kinetic energy (rolling) was converted into kinetic energy (falling) and how chemical stored energy was converted into chemical energy.

Help the students understand that in each instance energy was converted from one form to another and was neither created nor destroyed. Help the students think through the last question. What happened to all the energy at the end of the experiment? There was no more kinetic energy, everything had stopped. There was no more gravitational stored energy, the cards had fallen. There was no more chemical stored energy or chemical energy once the reaction was complete. What happened to the energy? It was converted into a form of energy that is hard to convert — like heat energy. The heat energy was released into the air.

V. Just For Fun

Allow the students to explore the use of different materials to perform individual energy experiments and then connecting the small experiments as steps in a series. Help them make observations about the conversion of energy from one form to another. Have them record their results.

Some ideas are provided in the *Laboratory Notebook*.

Experiment 14

Geology Every Day

Materials Needed

- colored pencils

Objectives

In this experiment students will explore their surroundings and observe how geology affects their daily lives.

The objectives of this lesson are:

- To encourage students to observe their surroundings.
- To help students explore the different aspects of geology (rock-part, air-part, water-part, and bio-part) and note how these aspects are interconnected.

Experiment

I. Think About It

Read this section of the *Laboratory Notebook* with your students.

Have the students think about where they live. Help them think about their local surroundings, noting the weather, types of wildlife, landforms, and any other features that stand out.

❶-❼ Have them answer the questions in this section. There are no right answers for these questions. Just allow the students to explore their own ideas about the geology of their surroundings.

II. Observe It

Read this section of the *Laboratory Notebook* with your students.

❶-❹ Have your students make a list of all the geological features they see in a day. They can make the list as they travel during the day, or they can take a walk outside and make observations.

They are directed to make lists that include the various types of geological aspects: rock-part, water-part, air-part, and bio-part Have them note features such as parks, trees, lakes, rivers, mountains, the weather, and any other features that stand out.

III. What Did You Discover?

Read the questions with your students.

❶-❹ Have the students answer the questions. These can be answered orally or in writing. Again, there are no right answers and their answers will depend on what they actually observed.

IV. Why?

Read this section of the *Laboratory Notebook* with your students.

Discuss any questions that might come up.

V. Just For Fun

Read this section of the *Laboratory Notebook* with your students.

Help the students think about the various aspects of geology they have explored in this experiment and what those features might be like on the Moon.

Have them draw their ideas and encourage them to use their imagination. There are no right or wrong ideas in this exercise.

Experiment 15

Mud Pies

Materials Needed

- small shovel or garden trowel
- small pail or plastic container
- measuring cup
- dirt that contains rocks (.25 liter [1 cup])
- 1 tall clear glass container (approx. size: .5 liter [2 cups])
- flour (60 ml [1/4 cup])
- water
- cake mix
- items needed to make the cake
- nuts, gumdrops, chocolate chips, and/or M&Ms

Objectives

In this experiment the students will explore how sedimentary rocks are made. The students will experiment with the sedimentation process by using a mixture of rocks, dirt, water, and flour.

The objectives of this lesson are for the students to:

- Observe the differences between rocks and dirt.
- Perform an experiment that illustrates the sedimentation process.

Experiment

I. Observe It

Read this section of the *Laboratory Notebook* with your students.

❶ Help your students collect a sample of dirt that contains rocks from their backyard, park, or some other place that will allow a sample to be taken.

❷ Use questions such as the following to help your students make good observations as they look through the dirt sample.

- *What does the dirt and rock mixture look like?*
- *Can you describe materials in this mixture that are different from each other?*
- *What do you think the different materials are?*

❸ Have your students use their hands to separate the rocks from the particles of soil. The rocks can be large or small.

❹ Guide your students' inquiry with the following questions:

- *What do the rocks look like?*
- *Can you describe their shape, color, or feel (texture)?*
- *What does the dirt look like?*
- *Can you describe the shape, color, or feel (texture) of the dirt?*
- *How can you tell the difference between a rock and dirt?*

❺ Help your students make a slurry of rocks, dirt, and water. There should be sufficient space in the clear glass container that a layer of water 5-8 centimeters (2-3 inches) thick can form on top of the rock and soil mixture.

❻ Guide your students' inquiry with the following questions:

- *What happens to the rocks when you swirl the mixture?*
- *What happens to the dirt when you swirl the mixture?*
- *What happens to the water when you swirl the mixture?*

❼ Have the students allow the mixture to settle and then help them record their observations.

II. Think About It

Read this section of the *Laboratory Notebook* with your students.

❶-❷ Have your students answer the questions. Have them base their answers on their actual observations and not on what they think should have happened.

Use the following questions to help your students think about why layers form.

- *Do you think rocks are heavier or lighter than dirt? Why or why not?*
- *Does the water make the rocks and dirt settle to the bottom more quickly or more slowly? Why or why not?*
- *What do you think would happen if you added oil instead of water to your mixture? Would the rocks settle more quickly or more slowly? Why or why not?*

Help your students see that layers are formed as heavier objects settle first and lighter objects settle last. The water slows down the settling process because water is thicker than air. Oil would slow it down even further since oil is thicker than water.

❸ Have the students add 60 ml (1/4 cup) of flour to the slurry and stir it. The flour will mix with the lighter dirt particles and may form a muddy-white layer when the mixture has settled.

Have the students record their observations.

III. What Did You Discover?

Read the questions with your students.

❶-❹ Discuss the questions in this section with the students. Their answers may vary since the answers are based on what they actually observed. There are no right or wrong answers to these questions.

IV. Why?

Read this section of the *Laboratory Notebook* with your students.

Explain to the students that this experiment shows how sedimentary rocks form. The word sediment comes from the Latin word *sedimentum* which means "to settle" and refers to solid matter that settles out of a fluid such as water or air. Sedimentary rocks begin to form when sediments settle and form layers as some sediments settle faster than others. When the layers of sediment are put under pressure, sedimentary rocks are formed.

What your students observed was exactly this process of sedimentation. The heavier rocks settled first, followed by the lighter rocks and dirt, leaving a layered sediment. Most sedimentary rocks are formed in this fashion.

V. Just For Fun

Read this section of the *Laboratory Notebook* with your students.

A great way to illustrate the process of sedimentation is to make a cake with layers of heavier nuts, gumdrops, chocolate chips, and/or M&Ms. Most recipes ask that you put flour on the additions to keep them from falling to the bottom. However, for this recipe do not use the flour so that students can observe how the heavier items settle to the bottom of the pan during the baking process.

Experiment 16

The Shape of Earth

Materials Needed

- baseball or similar hard-centered ball
- balloon
- water
- piece of string to tie balloon closed
- colored pencils

Optional

- funnel

Objectives

In this experiment students will explore how Earth gets its shape. Earth is almost spherical but bulges slightly at the equator. The central bulge is created by centrifugal force, an outward force caused by the rotation of Earth on its axis.

The objectives of this lesson are for students to:

- Observe the difference between a hard sphere (baseball) and a liquid-centered "sphere" (water balloon).
- Explore how the creation of centrifugal force by the Earth's rotation creates a central bulge.

Experiment

I. Think About It

Read this section of the *Laboratory Notebook* with your students.

Have the students think about the shape of Earth and why it is shaped like a slightly smashed ball. There is no right answer to this question. Allow the students to explore their own ideas about the cause of Earth's shape.

II. Observe It

Read this section of the *Laboratory Notebook* with your students.

❶ Have your students observe what happens when they spin a hard ball, such as a baseball, on the floor. Have them note whether the ball changes shape.

❷ Have them record their observations.

❸ Help the students fill a balloon with water and tie it closed. Putting the opening of the balloon over a garden hose works well, or you can use a funnel. Have the students observe what happens when they spin a fluid-centered ball, such as the water balloon, on the floor. Have them note whether or not the water balloon changes shape as it spins.

❹ Have them record their observations.

III. What Did You Discover?

Read this section of the *Laboratory Notebook* with your students.

❶-❹ Have the students answer the questions. These can be answered orally or in writing. Again, there are no right answers, and their answers will depend on what they actually observed.

IV. Why?

Read this section of the *Laboratory Notebook* with your students.

Discuss any questions that might come up.

V. Just For Fun

In this section students are introduced to thought experiments which are performed by using the imagination to think through the possibilities of a theory rather than by doing a physical experiment. Einstein was a famous scientist who used thought experiments to lead to new discoveries in physics.

Students are asked to imagine what it would be like if Earth were cube-shaped instead of spherical. Help spark your students' imagination with questions such as the following:

- *What do you think would happen if you were riding in a car and you came to an edge of the Earth?*
- *Would it make a difference if the edges of the Earth were sharp or if they were rounded?*
- *What would it be like if your house was near a corner of the Earth?*
- *What would the oceans be like?*
- *What would Earth look like if you were high up in an airplane?*

There are no right answers. Encourage students to use their imagination freely.

Have them record their ideas.

Experiment 17

Mud Volcanoes

Materials Needed

- 2 liters (8 cups) or more of dirt suitable for making mud pies
- 1.75 liters (7 cups) or more of water
- 15 milliliters (1 tablespoon) baking soda
- 15 milliliters (1 tablespoon) vinegar
- measuring cup
- measuring spoon
- 3 containers for mixing mud (about 1.75 liter [7 cups] size)
- spoon or other implement for mixing mud
- garden trowel
- bucket
- paper
- marking pen
- pencil
- colored pencils

Objectives

In this experiment students will make different mixtures of dirt and water to explore how the viscosity of lava determines the type of volcano formed.

The objectives of this lesson are for students to:

- Observe the differences between thick and thin mud mixtures.
- Observe how different mud mixtures will form different types of mounds and then compare these results to the way different types of volcanoes are formed.

Experiment

I. Think About It

Have the students think about the questions in this section. They can write their answers in the space provided or they can give their answers orally. There are no right or wrong answers to these questions.

Use questions such as the following to guide open inquiry.

- *When you pour syrup on your pancakes, what happens? Why do you think this happens?*
- *How fast does the syrup move? Why?*
- *How would the syrup act if you poured so much of it on your pancakes that it ran off the edge of the plate? Why?*
- *If you poured water on your pancakes, do you think it would be different from pouring syrup on them? Why or why not?*
- *How fast would the water move? Why?*
- *Is there a difference in how thick and thin liquids move? Why or why not?*
- *What do you think you would need to make a good mud pie?*

II. Observe It

Have the students dig up dirt that is suitable for making mud pies, or provide it for them. They will need at least 1.5 liters (6 cups) of dirt for this part of the experiment and .5 liter (2 cups) or more of dirt for the *Just For Fun* section.

Part I

Have the students follow the directions in the *Laboratory Workbook* to make 3 separate mud mixtures. Have them label each container of mud as **A**, **B**, or **C**.
A marking pen can be used or the students can come up with their own ideas for labeling.

If needed, have the students adjust the dirt:water ratio so that mixture **A** is like a thick paste, mixture **B** is somewhat thinner, and mixture **C** is more liquid.

[Parts II-IV: A table follows each set of questions in **Parts II-IV**. Have the students answer the questions by filling in the appropriate boxes in the table, or they may answer orally. Answers will vary and there are no right or wrong answers.]

Part II

Have the students observe the thickness of each mixture. Use questions such as the following to guide inquiry.

- *What happens when you add more water to the dirt? Why?*
- *Which mud mixture is easiest to mix? Hardest? Why?*
- *Do you think if you make a mud pie from each of the mixtures they will all look alike? Why or why not?*

Part III

Have the students select three separate areas in which to pour the three different mud mixtures. Each area should be labeled **A**, **B**, or **C** with a piece of paper or in some other way.

In general, students should observe that mixture **A** is thicker and more difficult to pour than liquid **C** and that mixture **A** will form a higher mound while mixture **C** will spread out the most.

Use questions such as the following to guide inquiry.

- *Do the different mixtures pour differently? In what ways?*
- *Do all the mixtures spread out in the same way after they are poured? Why or why not?*
- *Which mixture covers the largest area? Why?*

Part IV

In this part of the experiment students will add more layers of mud to each mound, letting the mud pie dry out each time before adding another layer of mud. They are to pour the same mixture in the same spot each time. They should observe that liquid **A** forms a more compact mound than liquid **B** or **C** and that liquid **A** takes fewer layers to build up the mound.

Use questions such as the following to guide inquiry.

- *Do the mixtures all make mounds of the same height with the same number of layers? Why or why not?*
- *Which mound is the widest? The highest, the middle height, or the lowest? Why?*
- *What do you observe about the relationship of height to width in the different mounds? Why do you think they are different?*

III. What Did You Discover?

Have the students answer the questions in this section. In general they should observe that the thicker mud of mixture **A** would make a cone volcano shape and the thinner mud of mixture **C** would make a shield volcano shape.

IV. Why?

This experiment helps students explore how different types of lava will form different types of volcanoes. Thick lava that does not flow very far away from the vent where it comes out of the Earth forms cone volcanoes, and thinner lava that can flow very long distances forms shield volcanoes.

V. Just For Fun

Have the students build a mock cone volcano using a thick mud mixture. After they form a mound with the mud, they can use a pencil to poke a hole down the center of the mound while the mud is still wet.

When the mud has dried, help the students measure and pour 15 milliliters (1 tablespoon) of baking soda and then 15 milliliters (1 tablespoon) of vinegar into the center of their volcano and observe what happens. Direct them to avoid looking directly into the volcano opening when they pour the vinegar in. They can record their observations by drawing them in the box provided.

Experiment 18

Observing the Stars

Materials Needed

- clear night sky
- colored pencils

Note: This experiment will take 6 days to complete.

Objectives

In this experiment, students will observe the stars, Moon, and planets on several different nights to determine their movement. Students will also explore how using the stars for navigation (celestial navigation) is possible.

The objectives of this lesson are:

- To observe changes in the position of stars, planets, and the Moon.
- To understand that it is possible to use the stars for navigation.

Experiment

I. Think About It

Read this section of the *Laboratory Notebook* with your students.

❶-❷ Have the students think about different ways they travel. Have them record their answers.

❸ Explain that *to navigate* means to plan and follow a route from one place to another. Have the students think about how a driver, pilot, or captain can find out which direction to travel. Discuss possible navigation tools such as:

- *compass*
- *GPS (Global Positioning System)*
- *radio navigation*
- *maps*

Have the students think about how using these tools enables modern people to travel to their destination. Have them record their answers.

❹ Help the students imagine what it would be like not to have any modern navigation tools.

Use the following questions to have the students explore the possibility of using the stars for navigation.

- *How easy or difficult would it be to use the stars to navigate?*
- *What happens when the sky is cloudy?*
- *Would traveling during the daylight be possible?*

II. Observe It

Read this section of the *Laboratory Notebook* with your students.

❶-❹ Have the students observe the night sky for six nights. Pick a single location (the backyard or front porch, for example), and observe the sky at the same time each night.

Help the students draw what they see. They do not have to draw every star. Try to help them find prominent stars and locate the same stars each night. Have them note if any of the stars have moved.

III. What Did You Discover?

Read the questions with your students.

❶-❹ Have the students answer the questions. These can be answered orally or in writing. There are no right answers, and their answers will depend on what they actually observed.

IV. Why?

Read this section of the *Laboratory Notebook* with your students.

Discuss any questions that might come up.

V. Just For Fun

Read this section of the *Laboratory Notebook* with your students.

Encourage the students to use their imagination to make up their own constellations.

Experiment 19

Earth in Space

Materials Needed

- basketball
- ping-pong ball
- flashlight
- empty toilet paper tube
- glue or tape
- scissors
- marking pen
- a dark room

Objectives

In this experiment students will use simple materials to explore how light from the Sun affects the Earth and the Moon.

The objectives of this lesson are:

- To have the students observe how the Sun illuminates the Earth and the Moon.
- To demonstrate lunar and solar eclipses and the seasons.

Experiment

I. Observe It

In this section students will make a model Earth and use it to explore the illumination of the Earth and the Moon by the Sun.

Read this section of the *Laboratory Notebook* with your students.

❶-❷ Here the students will make a model of the Earth. Help them cut out the continents from the page and paste or tape them on the basketball in the appropriate positions, with North America, South America, and Greenland on one side; and Australia, Africa, Europe, Russia, and Asia on the other side. Have students refer to a globe if needed.

Have the students mark the approximate location of the North and South Poles by taping or gluing small pieces of paper to the basketball or by using a marking pen.

❸ Help your students cut the toilet paper cylinder 2.5 cm (one inch) from the end. This will create a ring to hold the basketball. Have them place the basketball on the ring and slightly tilt the ball so the North Pole is pointed slightly to the side. This represents the tilt of Earth's axis which is about 23° from vertical.

❹ A flashlight will be used to model the Sun. Have the students turn off the room lights and shine the flashlight on the basketball from some distance away. Have them observe how the flashlight illuminates the basketball.

❺ Students can place the flashlight on the floor, or you can hold the flashlight for them. They will now slowly rotate the basketball in a counterclockwise direction to simulate the rotation of Earth on its axis.

Guide your students' inquiry with the following questions:

- *Does the light cover the whole basketball or just one side?*
  just one side
- *If the light is shining on Asia, is North America light or dark?*
  dark
- *As you rotate the ball, does the light on Asia stay the same?*
  No, it changes.
- *Are Asia and Russia illuminated at the same time?*
  yes
- *Are Asia and North America illuminated at the same time?*
  no
- *If it is light in Asia, do you think it will be day or night in Russia?*
  day
- *If it is dark in Asia, do you think it will be day or night in North America?*
  day
- *Do you think that if the light is shining on South America it will be day or night in North America?*
  day

Have the students record their observations.

❻ Now the students will model the Moon. Have your students take the ping-pong ball and place it between the basketball and the flashlight, some distance away from the basketball. They will have to hold the ping-pong ball with their fingers. Have them observe the shadow the ping-pong ball casts on the basketball when it is between the basketball and the flashlight. This represents a solar eclipse during which the Moon blocks sunlight from reaching a portion of the Earth.

Have your students move the ping-pong ball in a circle around the basketball. As the ping-pong ball goes behind the basketball, the basketball casts a shadow on the ping-pong ball. This represents a lunar eclipse when Earth's shadow falls on the Moon.

Guide student inquiry with the following questions:

- *What happens to the basketball when the ping-pong ball is between the flashlight and the basketball?*
  [This represents the Moon between the Sun and the Earth—a solar eclipse.]
  The basketball will have a round shadow on it created by the ping-pong ball.
- *What happens when the basketball is between the flashlight and the ping-pong ball?*
  [This represents the Earth between the Sun and the Moon—a lunar eclipse.]
  The ping-pong ball is in the shadow created by the basketball.

Have the students record their observations.

II. Think About It

Read this section of the *Laboratory Notebook* with your students.

❶-❸ Have the students answer the questions. Encourage them to answer in their own words. Suggested answers are shown below.

(Answers may vary.)

❶ Can you determine how day and night are created by the rotation of Earth?

As the Earth rotates, the Sun shines on different parts of the globe.

❷ Can you observe how a lunar eclipse forms (where Earth casts a shadow on the Moon)?

When the Earth blocks the Sun's light from the Moon, the Moon has a shadow on it from the Earth.

❸ Can you observe how a solar eclipse forms (where the Moon casts a shadow on Earth)?

When the Moon blocks the Sun's light from the Earth, the Moon's shadow falls on the Earth.

❹ *Using the basketball and flashlight, can you show how the seasons are created? Explain how you would do this.*

Help your students model how the orbiting of Earth around the Sun creates seasons. Have the students stand some distance away from the flashlight Sun, holding the basketball Earth and tilting it slightly as it was in the *Observe It* section. One pole should be pointed toward the flashlight. Students will need to keep the tilt of the basketball constant as they circle the flashlight.

Have them hold the basketball and walk around the flashlight in a counterclockwise direction. To simulate the orbit of Earth around the Sun, students will remain facing in the same direction as they circle the flashlight. With the flashlight on their left and the "pole" of the basketball pointing toward it, they will start walking forward, then to the left, then backward, to the right, and forward again, completing the circle. Have them notice how one pole is tilted toward the Sun on one side of the circle and the other pole is tilted toward the Sun on the opposite side of the circle.

Discuss with the students how the parts of the Earth that are tilted toward the Sun receive more heat energy from the Sun and the parts tilted away receive less. At different times of the year different parts of the Earth are tilted more toward the Sun.

III. What Did You Discover?

Read this section of the *Laboratory Notebook* with the students.

❶-❹ Discuss the questions in this section with the students. Have them record their answers. Since these answers are based on what the students actually observed, their answers may vary.

IV. Why?

Read this section of the *Laboratory Notebook* with your students.

Review with the students how in this experiment they used a flashlight to represent the Sun, a basketball for Earth, and a ping-pong ball for the Moon. Explain that by doing this they built a model and that scientists build models to help them understand how things work. With this model the students were able to explore how the Earth rotates, creating day and night; how the movements and positions of the Moon and Earth create lunar and solar eclipses; and how seasons occur. (Students will learn more about making models in a following experiment.)

V. Just For Fun

Read this section of the *Laboratory Notebook* with your students.

Help the students look at the map on the basketball and mark the approximate location of where they live and the location of the equator. Help them orient the basketball and flashlight for each part of this experiment.

Encourage students to use their imagination as they experiment with the ideas presented in this section.

There are no right answers here. The purpose of this exercise is to encourage students to make observations, explore ideas, and use their imagination.

Experiment 20

Seeing the Moon

Materials Needed

- colored pencils
- night sky

Note: This experiment will take two weeks to complete.

Objectives

In this experiment students will observe different phases of the Moon.

The objectives of this lesson are to have students:

- Practice making observations about a physical event that changes over time.
- Observe how their own observations may vary.

Experiment

I. Observe It

Read this section of the *Laboratory Notebook* with your students.

❶ Your students will be observing the movement, shape, and color of the Moon over fourteen nights. Pick a time during the evening to observe the Moon. It is recommended that the Moon be observed at the same time each night.

Guide your students' inquiry with the following questions.

- *What is the shape of the Moon?*
- *What is the color of the Moon?*
- *Can you observe any details in the Moon? Light or dark areas?*
- *Does the size of the Moon change?*

❷ Have the students record their observations each night.

II. Think About It

Read this section of the *Laboratory Notebook* with your students.

❶-❸ Have the students look at the various drawings or descriptions they have recorded. Have them use this information to answer the questions in as much detail as possible and in their own words.

III. What Did You Discover?

Read this section of the *Laboratory Notebook* with your students.

❶-❸ Discuss the questions in this section with the students. Have them record their answers. Answers may vary.

IV. Why?

Read this section of the *Laboratory Notebook* with your students.

Discuss with the students how the Moon appears to change shape, size, and color as it circles the Earth. Explain that the actual physical shape of the Moon does not change (the Moon itself doesn't actually grow larger or smaller or become full or half) but what changes is the way the Sun illuminates the Moon. Explain that as the Moon orbits the Earth, the Sun's light hits the Moon from different angles. The difference in how the Moon is illuminated by the Sun is what causes the Moon to appear to change shape.

Explain that when the Moon is closer to the horizon, it appears larger than when it is farther up in the sky. This happens because as the Moon nears the horizon, the atmosphere bends the light, making the Moon appear magnified. However, the physical size of the Moon has not changed.

V. Just For Fun

Read the text with your students.

Have your students observe any features of the Moon they find interesting. Can they see the "Man in the Moon?" Or "Jack and Jill?" Encourage them to use their imagination.

Have them draw what they observe and imagine. There are no right or wrong answers.

Experiment 21

Modeling the Planets

Materials Needed

- 8 Styrofoam craft balls
  Recommended sizes:
  1 - 10 cm (4 inch) ball
  1 - 7.5 cm (3 inch) ball
  2 - 5 cm (2 inch) balls
  2 - 4 cm (1.5 in) balls
  2 - 2.5 cm (1 in) balls
- Water-based craft paint
  Recommended colors:
  red
  blue
  green
  orange
  brown
- paint brush
- water in a container
- *Just For Fun* section— miscellaneous objects to use to represent planets (such as fruits, vegetables, candies, baking mixes)

Objectives

In this experiment students will create a model of each of the planets.

The objectives of this lesson are for students to:

- Learn how to construct models.
- Explore the advantages and limitations of creating models.

Experiment

I. Observe It

Read this section of the *Laboratory Notebook* with your students.

❶-❷ In this section have the students look up the relative size and colors of each planet, using the illustrations in their *Student Textbook* or another reference.

Guide your students' inquiry with the following questions.

- *Of the eight planets, which one is the largest?*
  Jupiter
- *Which planet is the smallest?*
  Mercury
- *Are there any planets that are similar in size?*
  Earth and Venus, Neptune and Uranus
- *What color is Earth?*
  blue-green
- *What color is Mars?*
  reddish brown
- *What color is Neptune?*
  blue
- *What color is Jupiter?*
  brown stripes

Have the students record their observations about what each planet looks like.

❸ Have students choose a Styrofoam ball to represent each planet. They will need to compare the relative sizes of the planets to the various sizes of Styrofoam balls. Let your students decide on the sizes even if they are "wrong." A very important part of science is creating a model and discovering how well the model fits reality.

❹ Have the students paint the Styrofoam balls according to the information they've collected.

II. Think About It

Read this section of the *Laboratory Notebook* with your students.

❶-❸ Have the students think about how they chose the Styrofoam ball that would represent a certain planet. In general, see if they understand that for an accurate model the largest planet would need the largest Styrofoam ball, and the smallest planet would need the smallest Styrofoam ball. Discuss why this is important for building this model.

III. What Did You Discover?

Read this section of the *Laboratory Notebook* with your students.

❶-❸ Have the students discuss what they observed during the process of model building. Answers will vary.

IV. Why?

Read this section of the *Laboratory Notebook* with your students.

Discuss the process of model building. Model building is an important part of science. Models give scientists a way to visualize things they cannot observe directly. Models can also give scientists a deeper understanding of how natural processes work.

Discuss the limitations of model building. Models are only models and not "reality." Just because a scientist builds a model does not mean that it is an accurate representation of reality.

Scientists build physical models, like the students did in this project, but scientists also build mental models of how things work. Mathematical explanations and scientific concepts are part of the "models" scientists create to help them understand how the world works.

V. Just For Fun

Read this section of the *Laboratory Notebook* with your students.

Help your students create planet models from other materials. Edible planets can be made from fruits, vegetables, or candy. Planet models could also be baked using cupcakes or bread, or inedible objects could be used. Encourage your students to use their imagination.

The *Laboratory Notebook* has a space for students to list ideas for making planet models, and on the following page it has a space for students to draw the completed model.

Experiment 22

Putting It All Together

Materials Needed

- ladybug or other small insect or animal to observe
- magnifying glass
- *Just For Fun* section: a second living thing or an object to observe

Objectives

In this experiment students will observe a living thing, such as a ladybug, and explore how each of the five scientific disciplines — chemistry, biology, physics, geology, and astronomy — are needed for learning about it.

The objectives of this lesson are for the students to:

- Think about how the five core scientific disciplines overlap and support learning about a living thing.
- To understand how the five core disciplines are interconnected.

Experiment

I. Think About It

Read this section of the *Laboratory Notebook* with your students.

❶-❻ Have the students think about how the five scientific subjects are connected to each other. Then have them record their answers.

There are no right answers for this question. Allow the students to explore their own ideas.

II. Observe It

Read this section of the *Laboratory Notebook* with your students.

❶ Have your students observe a living thing such as a ladybug. If ladybugs are not available where you live, they can choose another living thing.

❷ As they answer the questions, have the students fill in the blanks with the name of the living thing they observed. Help your students think about each question and whether or not the question is asking about chemistry, physics, biology, geology, or astronomy. Here are some suggested answers. However, your students' answers may vary. If there are some questions for which they are unable to observe the answers, they can record their idea of what they think the answer would be.

- *What do ladybugs eat?*
  small bugs called aphids (biology)
- *How do ladybugs digest food?*
  chopping with mandibles (physics) then with chemicals inside the abdomen (chemistry)
- *How do ladybugs move?*
  flying (physics, biology)

- *Where do ladybugs live?*
  in many types of climates and wetlands (geology)
- *How long do ladybugs live?*
  several months (biology)
- *How big are ladybugs?*
  1-10 mm (physics)
- *How much does a ladybug weight?*
  1 gram (physics)
- *What do ladybugs breathe?*
  oxygen (chemistry)
- *Would ladybugs be found on the Moon? Why or why not?*
  No. The Moon does not have an oxygen atmosphere. (geology and astronomy)

III. What Did You Discover?

Read the questions with your students.

❶-❹ As they answer the questions, have the students fill in the blanks with the name of the living thing they observed. The questions can be answered orally or in writing. Again, there are no right answers, and their answers will depend on what they actually observed. If there are some questions for which they are unable to observe the answers, they can record their idea of what they think the answer would be.

IV. Why?

Read this section of the *Laboratory Notebook* with your students.

Help the students see how science is a combination of all five core disciplines. Discuss any questions that might come up.

V. Just For Fun

Read this section of the *Laboratory Notebook* with your students.

Have the students select a living thing or object to think about. Have them record the name of the item in their *Laboratory Notebook.*

Ask the students questions to help them think of some facts about the thing they've chosen. Help them think about the branch of science that would help them learn about the fact. Have them record their ideas. There are no right answers to this exercise.